GodLight

GodLight

Possibilities from the Intersections of Science and Spirituality

Bart Barthelemy

Published by **Bart Barthelemy** in partnership with

Quantum Wisdom Press

www.quantumwisdompress.com

GodLight

Copyright © Bart Barthelemy

All rights reserved. No part of this book may be used or reproduced in any manner whatsoever or any information storage and retrieval system without written permission except in the case of brief quotations embodied in critical articles or reviews and with attribution to the author.

This book is a work of non-fiction.

For information visit www.godlightbook.com

ISBN: 979-8-9859838-1-4 (pb)

First Edition: (2022)

Dedication

To Gretchen, who is closest to the Light.

Introduction	1
Science and Spirituality	3
Prologue	5
GodLight	9
The Light of the Universe	11
The Light of the World	13
Cause of the Universe/Creator of the World	15
Cause of the Universe	15
Creator of the World	17
Possibilities	18
Big Bang/Creation	19
Big Bang	19
Creation	21
Possibilities	23
Physical Light/Spiritual Light	25
Physical Light	25
Spiritual Light	27
Possibilities	30
SpaceTime/Space & Time	31
SpaceTime	31
Space and Time	33
Possibilities	34
Dark Matter/Mystery	35
Dark Matter	35
Mystery	37
Possibilities	38
Black Holes/Hell	39
Black Holes	39

Hell	41
Possibilities	42
The Universe/Heaven	43
The Universe	43
Heaven	45
Possibilities	45
Infinity/Eternity	47
Infinity	47
Eternity	49
Possibilities	50
Planets/Earth	51
Planets	51
Earth	53
Possibilities	54
Evolution/Creation	57
Evolution	57
Creation	59
Possibilities	61
Humans/Souls	63
Humans	63
Souls	65
Possibilities	66
Opportunities	67
Epilogue	69
About the Author	71

Foreword

Why are we here? What is our purpose? Who is God? How do we know that God exists? How was the universe created? Seeking answers to these questions can be a lifelong journey for so many of us. Finding meaning and inner peace are at the core of our search for answers. We devote so much of our lives to managing uncertainty and gaining comfort with it. Both science and spirituality help us manage uncertainty in life. While science reduces uncertainty, spirituality helps us make peace with it. GodLight offers an array of possibilities for answering these and other questions as Bart Barthelemy uncovers patterns of reality in scientific studies and spiritual realms.

What began as Bart's revelation as a five-year-old Catholic student became a noble hypothesis and a lifetime endeavor to see God as Light. This marked the beginning of his spiritual journey and scientific inquiry as well as his vision for GodLight.

In this text, Bart takes us through his lifelong spiritual exploration and scientific discovery. He discovers insights, commonalities, and collaboration between the scientific and spiritual paradigms. He notices how often the sacred texts of all the major world religions reference light. He explains how scientific examination of light has led to breakthrough discoveries about the interaction of matter and energy as well as space and time.

While science supplies answers and breakthroughs, there is still much of the universe that remains unknown. Bart reveals how unknown areas of science could be explored from a new spiritual perspective while a faith-based view could be explained by science. He enlightens us to reflect on the overlap of science and spirituality.

Bart diligently strives to make scientific concepts comprehensible. While many physics concepts described are incredibly complex, he makes them understandable without diluting or oversimplifying them. His explanations and illustrations facilitate our understanding of the complicated ideas. He references physics, chemistry, biology, mathematics, philosophy, psychology, religion, and spirituality.

While his multi-disciplinary approach reflects his academic rigor and resonates with scientists, Bart desires to educate and inspire all. GodLight may serve as another means to counteract false binary thinking in a world that promotes duality. We do not have to declare our allegiance to science while forgoing a spiritual life. Science and spirituality can co-exist in our mind, heart, and soul.

Bart summarizes GodLight with this inspiring quote:

"If believers see Light (or GodLight) as the fundamental element of everything in the universe, then understanding, collaboration, mutuality, respect, and universal love may be a more reasonable possibility."

GodLight can bring us closer to interpersonal harmony and universal connection. Experiencing God as Light can help solidify our understanding of the universe while offering us contentment as we explore the vast unknown.

—**CARL GAERTNER**, Applied Knowledge Management

Preface

The purpose of this book is to stimulate curiosity and invite new questions about the intersections of science and spirituality. Each chapter focuses on a major topic and is divided into three sections: the first section covers the topic from a scientific perspective; the second section covers the topic from a spiritual perspective; and the third or "Possibilities" section draws attention to the intersections of these two perspectives.

GodLight is best enjoyed in a contemplative state with time between chapters to consider the "Possibilities." Rather than drawing finite or absolute conclusions, GodLight is designed to invite questions and stimulate curiosity. Over time, our perspectives of these universal topics may evolve. Given that, we invite you to revisit the content from time to time and see if any new insights emerge. As Marcel Proust wrote, "The real voyage of discovery consists, not in seeking new landscapes, but in having new eyes."

Acknowledgements

I have lived a long and happy life thanks to my wonderful family, my many professional associates and all of my personal colleagues and friends. There are too many individuals to name, so I will acknowledge all of you collectively and spare myself any embarrassment by forgetting to mention any of you. Know that I appreciate all of you. Every person that I have encountered in life has led me to writing GodLight in one way or another. So to all, thank you.

This book is the wrap-up of my life. While I hope that it does get read, I would be really happy if my words inspired just one other person to see God as the Light of the World. So writing and publishing GodLight seemed necessary and thank God that a few kind friends helped with that. Olivia Parr-Rud, Carl Gaertner and Marta Reis have been tremendously supportive and wonderfully helpful. Without their encouragement and support, my words would have remained in the cloud and not seen the light of day. Emily Riley, Candace Dalmagne-Rouge, Linda Cindrich and Kathy Hollingsworth have provided great insights and comments and inspired me to make this happen. Finally, I have to thank Marge, who has put up with me and all my crazy adventures for the past sixty five years. While we are as different as can be, I know that it is GodLight that has made our marriage successful and my life balanced.

Introduction

This book is about the intersections of science and spirituality. As a scientist, I have always been fascinated by the important role that light plays in many aspects of science. But the motivation to write a book was sparked by the repeated number of spiritual learnings and experiences involving light that I have had in recent years. Because of this, I began to explore the significance of light in both disciplines. The result of this study led to the concept of GodLight, the Light of God, that I believe is foundational to both science and spirituality.

Science and Spirituality

Science is the study of the nature and behavior of natural things and the knowledge we obtain about them. Most readers of this book will likely have a general knowledge of science and how our world works. The fundamental concepts in science are learned during our early education and greatly enhanced by the tremendous amount of information that is available today on the internet and television. Our current understanding is that we have more scientific knowledge today than any prior generation. Some of us who call ourselves scientists have more specific knowledge in certain areas of science. But, in general, we are all aware of the key concepts in science, such as the Big Bang, the expansion of the universe, galaxies, black holes, our solar system, the planets around our sun, the Earth, the evolution of human beings and the biological complexity of men and women. These are the aspects of science that will be discussed in this book.

Spirituality is the belief that there is something greater than ourselves, something more to being human than sensory experience, and that the greater whole of which we are all a part is cosmic or divine in nature. Most of us believe this and, while not identical, we often equate spirituality to religion. At their core, most religions agree that there is a deity called God and that human beings need to respect and love each other. There are many other specifics that separate the various religions. But there are also an amazing number of commonalities. These are the aspects of spirituality that will be discussed in this book.

Each section will look for intersections and overlaps between science and spirituality in a particular area. There is no attempt to conform science to spirituality or vice versa. The focus is to simply examine the intersections to see if they suggest possibilities for a better understanding of each area. If so, these possibilities could lead to opportunities for any of us to be more understanding, more respectful, more collaborative, and more loving. Even though there are still great mysteries to be studied and understood in both science and spirituality, what we now know and believe in each of these areas is more overlapping and connectable than ever before.

As you read each section, ponder the marvels and mysteries of both science and spirituality. Let your mind and your imagination be open to possibilities.

For centuries, human beings have generally separated these two disciplines and often struggled with the conflicts between science and spirituality. This book is an invitation to look at the commonalities and confluence of these two areas so that there is more agreement than separation, more harmony than conflict, and more understanding than argument. Hopefully, this leads to a future that improves upon our past.

Prologue

When I was five years old, I was sure that God was a huge light that shined on everything. Perhaps that was because the church that we attended every Sunday had this magnificent, three dimensional montage in the back of the altar that showed brilliant beams of light streaming through clouds and storms onto an array of angels, saints, ordinary people, animals, and plants. So when I entered St. Anthony Elementary School a year later, my answer to the nun who asked the question, "Who is God?" was simply "Light".

I spent the next eight years learning that "Light" was not the correct answer. I realized that God is so many other things, like part of a mysterious Trinity, a Father with a virgin mother to an Earthly god-man named Jesus, a vengeful deity who punishes bad people, a loving old-man who forgives our sins, the victor in the continuous fight with the devil and bad angels, the king of heaven who is surrounded on his throne by good angels, and so many more characters, adjectives, nouns, and verbs. While my thinking has become more sophisticated over time, those images have stuck with me all my life.

After finishing elementary school, I put aside my religious education except for going to church regularly and occasionally trying to understand why there were so many different religions and philosophies about God. Because I excelled in science, I went to MIT, got a couple of degrees in science and engineering, and later added a PhD from Ohio State. With that education and my natural interest in science, I decided to spend my entire life in that professional arena, mostly helping government organizations with science, technology, and innovation. At one time, I really was a rocket scientist!

Most of my life has been in the scientific world. In that arena, the truth has to be verified consistently by facts. It has been an interesting journey, filled with constantly increasing knowledge about our world, how it works, and our place in it. Because of the tremendous advances in information technology over the past twenty-five years, we are able to learn whatever we want to know about the universe, the galaxies and suns, the planets and Earth, humans and mankind as well as everything from fundamental matter to highly complex systems. This enables us to formulate laws that help us understand and predict many things on our planet. However, in the scientific world, there are no angels, devils, miracles, holy spirits, resurrections of the dead, heaven, or hell. And yet, almost everyone that I have met on my journey through life, including myself, accept both a scientific universe and a spiritual world that co-exist in both time and space.

Most of my colleagues and friends believe in two separate paradigms that govern their world: spiritual and scientific. Some colleagues are so wedded to their religious beliefs that they make up false histories of our world that are simply not true. Unless all of "proven science" is wrong, our world did not begin four thousand years ago. Conversely, it's hard to believe that the universe started from nothing. But most of us take it as acceptable to embrace two different views of our world, the scientific upon which all of our existence is based and the spiritual which is based on myths, ancient stories, miracles and scientifically unsubstantiated "truths".

At this time when our world needs to come together more than ever, it seems reasonable to look for insights, commonalities, and collaboration between these two paradigms. Both religion and science are paradigms which bifurcate from the other looking simultaneously at our world as a multidimensional collage versus an image. A closer examination reveals that there is much more overlap than we acknowledge. Personally, my childish answer doesn't seem as silly from my current vantage point and I want to explore it again.

I still believe that light plays a major role in everything in the universe, including our planet and our species. Encouragingly, some of the early as well as current philosophers, scientists and religious leaders believe that light is a, if not, the fundamental nature of God. Could it be that the light of our world is the light of the universe? Could everything in the universe be God's Light, GodLight? And could it be that we and everyone else are also GodLight?

These are questions that will be examined as we unfold the various elements and intersections of science and spirituality. While there certainly could be many more intersections than can be examined, hopefully this work will stimulate more inquiries and explorations.

GodLight

In science, a theory must be supported by irrefutable and repeatable evidence before it is considered valid and useable. As science progressed over the past millennia, more and more evidence became available to prove the validity of certain models. Those that were validated became universal laws. The areas where there is insufficient evidence for validation became unknowns for further study

In spirituality, a religion or philosophy can be asserted as true even when there is insufficient evidence to prove that every aspect of that belief is valid. Those areas usually are treated as mysteries to be accepted in faith. While both science and spirituality govern our lives, for many they coexist as parallel domains that remain separate. But what if they did overlap? Could the unknowns of science be explored through spirituality? Could the mysteries of spirituality have some scientific basis to move them beyond faith?

GodLight is the light of God. If that light governs both the scientific and spiritual domains in which we live, then examining each could yield insights for both domains. Areas of science which are still unknown could be explored from a new perspective. And areas of spirituality which are based on faith could become more explainable. In each domain, there are areas that we know and areas that we don't know.

But are there areas in either domain that could provide insights or knowledge in the other domain? Breakthroughs in any specific area usually occur when a problem is explored from a very different perspective. One of Einstein's famous quotes is "We cannot solve our problems with the same thinking we used when we created them." If there is overlap, then the unknowns of science and the mysteries of spirituality should profit by examining them from the opposite perspective. If Godlight is the overlap of the two domains, could it shed new light on the many unanswered questions in both domains?

The Light of the Universe

For most of us, we take the physical world for granted. Tomorrow should be a lot like today, as was yesterday. Sure, there might be life changing events that could happen today, but generally these will be infrequent.

We assume that the sun will come up in the morning, light up our day, and go down in the evening. Even with fog, overcast and rain, the sun is still shining above the clouds and heating up our world. At night, the moon will reflect the light of the sun and depending on the season and the day, will provide some light. Above our sun and moon, there are stars shining in the heavens which at least provide us with light to find our way. Even more subtle and unnoticeable is radiation from distant galaxies and our Milky Way that falls on our planet at different times. As we now know, there is also an extremely small amount of microwave background radiation that was generated at the beginning of the universe and continues to bathe all of the universe, including our planet.

All of this, plus all of the man-made sources of light, are what we assume light up our world. Even the man-made sources use a variety of fuel, like coal, oil, wood, and gas that was generated by the external radiation from our sun at some time in the past. In essence, everything that provides the apparent light of our world comes from our sun and other suns and stars in our universe, whether generated at the Big Bang or later in the evolution of our universe.

For centuries, human beings worshipped the sun and gods they associated with the sun. Most prominent were the ancient Egyptians, who worshiped Ra, the Sun God, and created many monuments and structures devoted to Ra. Since the wellbeing of many tribes was also tied to their use of the sun's light and warmth, the sun took on a prominent role in the lives of many early human beings.

While light from the sun and stars was obvious, scientific discoveries in the early twentieth century led to a new kind of light. At the microscopic level, all of universe is made up of some form of light. The data from Einstein's experiments with photoelectric materials could only be explained if light is a fundamental aspect of everything in nature. At the quantum level, everything in the universe is both wave and particle. Light can be reduced to wave packets that appear in discrete quanta called photons. These wave packets are fundamental to all material in the universe. This led to Einstein's breakthrough formulations of the general and special theories of relativity. Another outcome of his research was the most famous equation in the universe, $E=mc^2$. Again, c, the speed of light, is a universal constant which ties matter to its equivalent energy. Even in science, there is more to light than meets the eye.

The Light of the World

Based on several studies published in 2020, approximately seven billion people on Earth (about 90% of the total population) believe that some type of force or entity exists that is beyond and/or within the physical nature of their lives. Often, this force or entity is called God. From there, any agreement about the nature of God diverges widely. At the most basic level, some believe that there exists a number of gods which individually govern a variety of natural phenomena like the sun and the wind and various animals. On the other end of the spectrum are the current monotheistic religious beliefs of a precise God that is based on voluminous writings that were created millennia ago. While there are not seven billion descriptions of God, there are certainly thousands of these descriptions and each is quite distinct. Interestingly, almost all include a God or gods that provides the light and energy necessary for the existence and sustenance of their followers.

From cave paintings, we believe that early man first thought of God as connected to nature and the mysterious forces that impacted their lives. Even something as predictable as the rising sun in the morning and the changes in seasons were still unexplainable to ancient humans. In retrospect, the creation of a higher power that controlled the sun and the seasons, which in turn impacted almost all aspects of their existence, is not surprising. While not the only God-like construct, the sun is consistently depicted in many archeological findings as a higher power. More recently, the Egyptian dynasties from 4000 BCE to 1000 BCE honored the sun and worshipped a variety of gods and goddesses that were connected to the sun.

Before 1000 BCE, most people described God through myth, poetry, dance, music, fertility, and nature. Although it was a violent world focused on survival, people still recognized that they belonged to something cosmic and meaningful. From artifacts and cave paintings, there is evidence that they believed that their world was enchanted and that the "supernatural" was everywhere. Around 500 BCE, many forms of spirituality emerged from Eastern sages, Jewish prophets, and Greek philosophers, eventually coalescing worldwide and providing the foundations for all our world's religions and major philosophies.

In the East, spirituality often took the form of the holistic thinking that is found in Hinduism, Taoism, and Buddhism. This holistic thinking allowed people to experience forms of participation with reality, themselves, and the divine. In the West, the Greek philosophers gave us a mediated participation through thought, reason, and philosophy. At the same time, many mystics seemed to enjoy real participation, even though it was usually seen as a very narrow gate available to only a few.

For many of the people that lived in Israel, there was a dramatic realization of an intimate union and group participation with God. They recognized the individually enlightened people like Moses or Isaiah. But they did something more. The notion of participation was widened to the Jewish group and beyond. God was saving their people as a whole. Participation was historical and social, not just individual. It is amazing that we have forgotten or ignored this. Instead, we make salvation all about private individuals going to heaven or hell. This is certainly a regression from the historical, collective, and even cosmic notion of salvation taught in the Bible.

Both the Hebrew scriptures and the Judaistic experience itself created a matrix into which a new realization could be communicated. Jesus offered us full and final participation in his own very holistic teaching. This allowed Jesus to speak of true union at all levels: with oneself, with neighbors, with outsiders, with enemies, with nature, and—through all of these—with the Divine.

Cause of the Universe/Creator of the World

Cause of the Universe

In most of the published scientific literature on the subject, the universe is defined as "all of space and time and their contents including planets, stars, galaxies, and all forms of matter and energy." The Big Bang theory is the prevailing cosmological description of the development of our universe. According to this theory, space and time emerged together about 13.8 billion years ago and the universe has been expanding ever since. While the spatial size of the entire universe is unknown, cosmic inflation indicates that it has a minimum diameter of 93 billion light years in the present day. The universe is often defined as "the totality of existence", or everything that exists, everything that has existed and everything that will exist.

For at least several hundred years, science has purposely avoided considering God as the cause of the universe. From a scientific perspective, this is a reasonable position since science must be based on measurable facts, clear models, experiential evidence and repeatable experimentation. Since God doesn't fit nicely into these categories, God cannot be scientifically accepted as the cause of the universe. However, the universal law of the conservation of energy suggests that something extremely powerful and energetic must have existed before the Big Bang.

Many noted scientists have postulated a variety of causes and descriptions of the universe, such as multiuniverses or multiverse, spontaneous creation, wormhole connections across spacetime, string theory versus particles, etc. Since science cannot observe or measure anything before the Big Bang, it is impossible to prove what happened before the accepted Big Bang singularity. However, studies continue to examine different descriptions of the universe in the hope of finding a single acceptable theory of everything. There are many competing hypotheses about what, if anything, preceded the Big Bang. Many physicists and philosophers refuse to speculate, doubting that information about prior states will ever be accessible. Based on the above, it seems reasonable to assume that, at this moment, science does not have a satisfactory answer as to what caused the universe.

Some speculative theories have proposed that our universe is but one of a set of disconnected universes, collectively labeled as the multiverse. If these theories are someday verified by empirical evidence from far-reaching telescopes, then we may be able to understand more about our world. However, the likelihood of this is very small since we still know so little about our own universe.

Interestingly, the universe appears to have much more matter than anti-matter. Since matter and antimatter, if equally produced at the Big Bang, would have completely annihilated all matter, the imbalance between matter and antimatter is partially responsible for the existence of all the matter that exists today. Another theory of how the universe started was offered by Stephen Hawking. Asking what happened before the Big Bang, Hawking said that was like asking "what is a mile north of the North Pole? It is not any place, or any time."

Two other approaches, that are being studied to explain the universe, are string theory and wormholes. String theory characterizes particles as tiny microscopically vibrating strings. Wormhole theory is a model which links disparate points in the universe. It can be visualized as a tunnel with two ends at separate points in spacetime. Whether wormholes actually exist remains to be seen.

Creator of the World

In contrast to the scientific community, most spiritual and religious practitioners do not disagree on what or who created the universe. The universe and our world was created by God or is God. However, there are multiple descriptions of God, not only between but even within the many past and current philosophies and religions. Theistic religions like Christianity, Islam, and Judaism believe that God is both the creator and sustainer of the universe. Polytheistic religions, such as Hinduism believe in many gods that created and sustain the universe. Deism, the belief in God but not religion, believe that God is the creator, but not the sustainer of the universe. Pantheistic philosophies, like Buddhism, identify the universe with a God that is absolute and transcendent. But this world, being merely its manifestation, is fragmented and imperfect. Nevertheless, most religious and spiritual people alive today firmly believe that God or Gods created our world.

Possibilities

In this area, the contrast between science and spirituality is very clear. In science, the cause of the universe is unknown or there is no cause. In almost all religions and spiritual practices, the creator of our world is God.

Interestingly, many scientists are spiritual and religious. Most believe that God did create the universe, but they cannot prove it. Generally, that is the belief that most of our world accepts, a universe that was created by a mysterious God. Many religions characterize that God in very specific ways, none of which can be proven and must be accepted in faith. However, if we are to believe the science that has served us so well in explaining the laws and behavior of the universe, God must be at least as powerful and energetic as the universe itself. If nothing else, the scientific law of conservation of energy would lead to that conclusion.

While not necessarily equal to the universe, it's reasonable to assume that there might be some similarities between God and the universe. If so, the light of the universe might be very similar, if not the same, as the light of God. Could that provide some insights into the nature of God or the mysteries of the universe? Looking deeper into the scientific and spiritual paradigms of cause and creation might provide more possibilities to ponder. Looking even deeper, the possibility that the universe is some reflection of its Creator is even more intriguing!

Big Bang/Creation

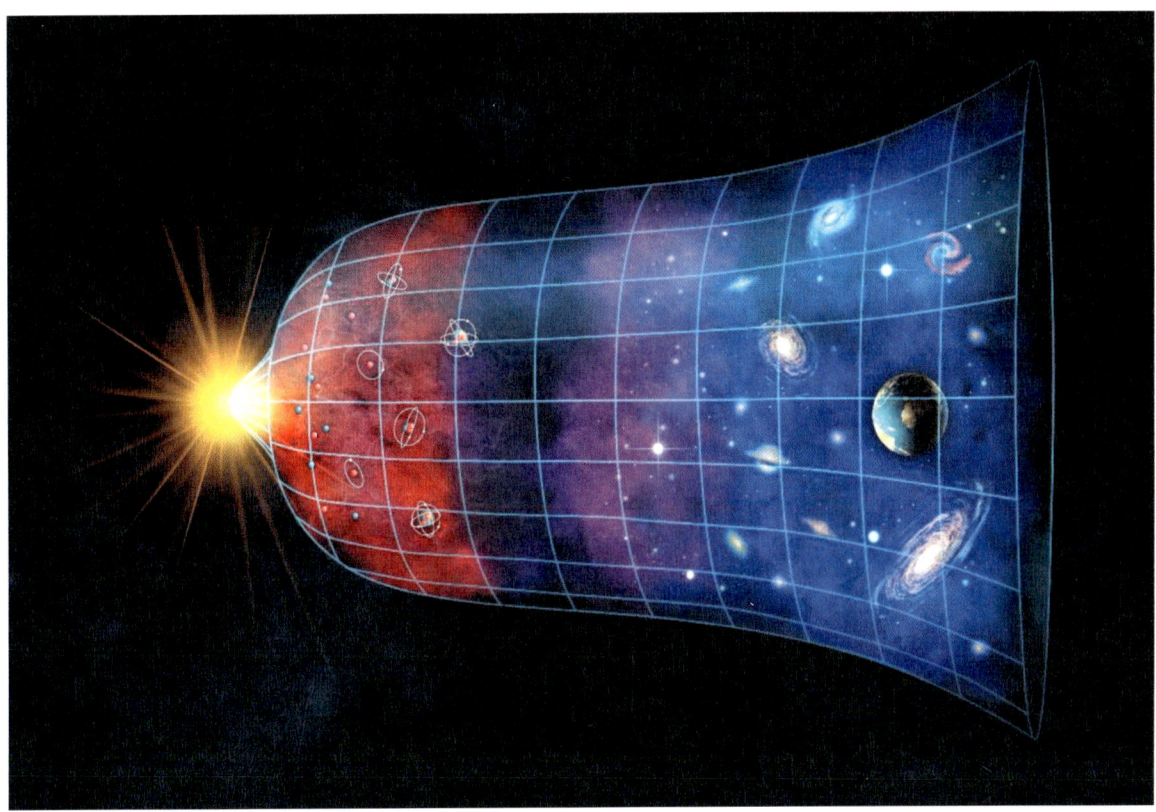

Big Bang

The Big Bang singularity is generally accepted by most scientists as the birth of the universe. According to theoretical models based on the general theory of relativity as well as a limited amount of empirical evidence, the energy and matter initially present in the singularity became less dense very quickly and the universe began to expand. This expansion, called the inflationary epoch, occurred extremely fast, in less than 10 to the minus 32 (10^{-32}) seconds. With expansion, the universe gradually cooled and continued to expand, allowing the first subatomic particles and simple atoms to form. Dark matter gradually gathered, forming a foam-like structure of filaments and voids under the influence of gravity. Giant clouds of hydrogen and helium were gradually drawn to the places where darkmatter was most dense, forming the first galaxies, stars and everything else seen today.

From studying the movement of galaxies, we now know that the universe contains much more matter and energy than is accounted for by visible objects; stars, galaxies, nebulas and interstellar gas. This unseen matter is known as dark matter because, while there is a wide range of indirect evidence that it exists, it has not yet been detected directly. The most widely accepted model of the universe suggests that about 69% of the energy in the universe is dark energy and about 26% is dark matter. Ordinary matter is therefore only about 5% of the physical universe with stars, planets and visible gas clouds only comprising about 6% of the ordinary matter.

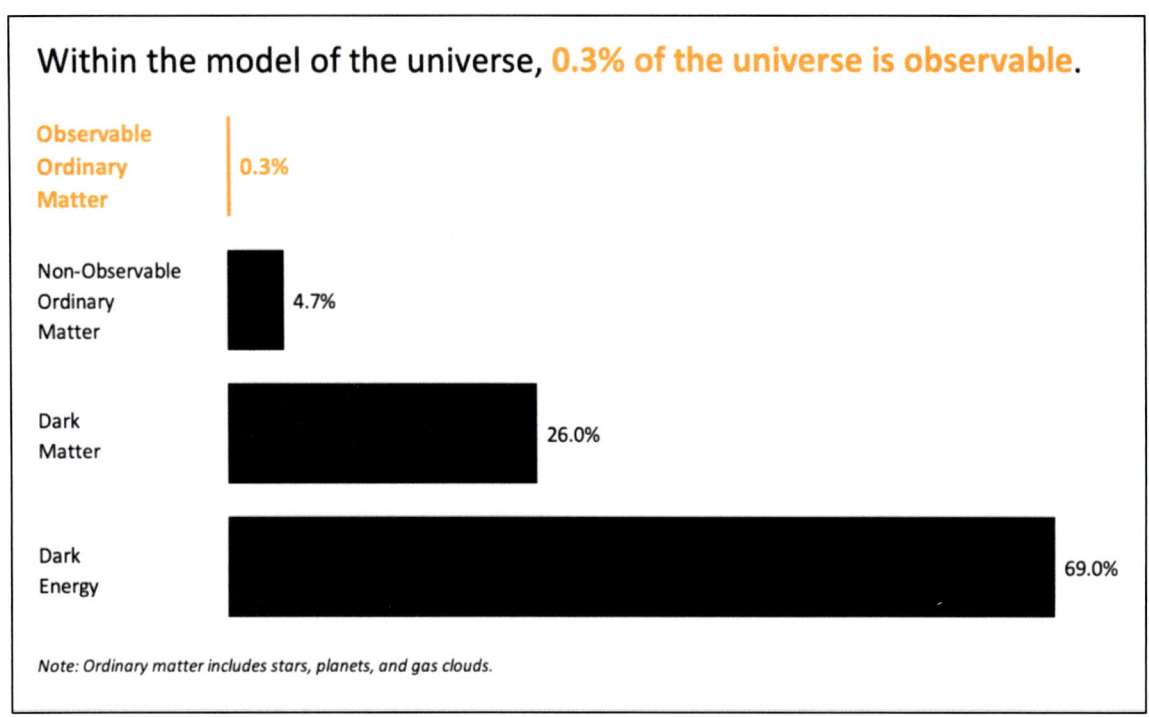

Amazingly, less than 0.3% of the universe is observable and more than 95% is dark and unknown. Despite centuries of study by some of the most brilliant minds of our world using the most sophisticated scientific instruments observing from space and on the Earth, we really know very little about the past as well as the current state of the overall universe.

Creation

Most of the major philosophies and religions have a creation story.

Hindus believe that the universe was created by Brahma, the creator who made the universe out of himself. After Brahma created our world, Vishnu is the god who preserves our world and all human beings. And as part of the cycle of birth, life and death, Shiva, another god, will ultimately destroy the universe.

For Christians, Genesis is the story of creation. Genesis states that, "In the beginning when God created the heavens and the Earth, the Earth was a formless void and darkness covered the face of the deep, while a wind from God swept over the face of the water and God then said, "Let there be light;" and there was light".

Genesis goes on to say, "On the second day, God created the sky, stars and suns, On the third day, God created dry land, seas, plants and trees, On the fourth day, God created the sun and the moon, On the fifth day, God created the creatures that live in the sea and creatures that fly, On the sixth day, God created the animals that live on the land and finally humans, made in the image of God, and on the seventh day, God finished his work of creation and rested."

While very few Christians believe that the length of the days referred to in Genesis are the same as our current 24 hour days, most do accept the sequence from light to sky to sun to creatures and finally to humans, as believable.

Actually, this sequence is strongly supported by the accepted evolution theories of the universe, the galaxies, the suns, the planets. In the context of this book, a most interesting aspect of the Christian Genesis creation story is that it begins with God saying "Let there be light".

The Judaic creation story has several versions that are slightly different from the Christian version. Many Jews accept the Biblical Genesis story described above. Again, the emphasis is that Light was the first thing that God created and everything else followed.

The Islamic Creation story comes from the Qur'an, which states "Allah created the heavens and the Earth, and all that is between them, in six days". While this is similar to the account related in the Bible, there are some distinctions.

Muslims interpret the description of a "six day" creation as six distinct periods or eons. The length of these periods is not precisely defined, nor are the specific developments that took place during each period. After completing the Creation, the Qur'an describes that Allah "settled Himself upon the Throne". But Allah is never "done" with His work, because the process of creation is ongoing. Each new child who is born, every seed that sprouts into a sapling, every new species that appears on Earth, is part of the ongoing process of Allah's creation.

There are scientific explanations for the existence of the Earth within Buddhism, but they are not tied to God. Most Buddhists do not believe in the concept of God.

Regardless of their religious or spiritual beliefs, most human beings believe that something created our world, generally describing that something as God. As we look around us, there is always something that preceded what we see now. The arrow of time is always forward. So the past led to the present which leads to the future. It's hard for most of us to conceive of something coming from nothing.

Possibilities

The exact nature of the Big Bang, both at the every instant that it occurred as well as during the seconds, hours, days, and years of the expansion, is an ongoing work of science. However, most scientists agree that there was a singularity with a tremendous amount of energy that occurred at that instant.

Similarly, most religions believe that there was a moment when God created the universe. The moment is when God said, "Let there be light". While scientists still debate exactly what happened after that singularity, the most direct evidence of the Big Bang is the Cosmic Background Radiation that is now observed throughout the universe.

The Cosmic Background Radiation is electromagnetic radiation, a form of light, from the Big Bang. This electromagnetic radiation depends on the region of the light spectrum that is observed and the most studied component is the Cosmic Radiation Background—light that has freely streamed from the beginning of the universe until today.

It appears to me that there is some overlap between the scientific and religious models of the beginning of the universe and our world. Light played a prominent role in both cases. Light in the physical world is electromagnetic radiation that was present at the beginning of the universe and continues to be seen today. Light in the spiritual world is how God started the universe and continues to be an important part of spirituality today. While the nature of these two "Lights" may be different, they play a major role not only during creation of the universe and our world but also in the evolution of both.

Physical Light/Spiritual Light

Physical Light

Light is electromagnetic radiation that can be described by its wavelength or its frequency. Light occurs over an extremely wide range of wavelengths from gamma rays with wavelengths less than 10 trillionth of a meter to radio waves with wavelengths in meters. Within that broad spectrum, wavelengths visible to humans, which are also generally referred to as light even though they are a small part of the entire electromagnetic spectrum of light, occupy a very narrow band, from about 700 nanometers for red light down to about 400 nanometers for violet light. The spectral regions near the visible band are commonly designated as infrared at the red end and ultraviolet at the violet end.

In the physical world, the broad spectrum of light has allowed human beings to use light from radio waves to communicate, light years (the distance that light travels in one year) to measure huge distances, infrared and microwave light to heat objects, and visible light to see objects. While seeing the light with our eyes is restricted to light in the visible wavelength regime, a broader range of light can be felt by the human body when the energy of light is absorbed by the body and produces heat. Sunbathers know that ultraviolet light can be dangerous to their skin after long exposures, while the infrared light in a microwave heater is absorbed by liquids and solids as it heats them.

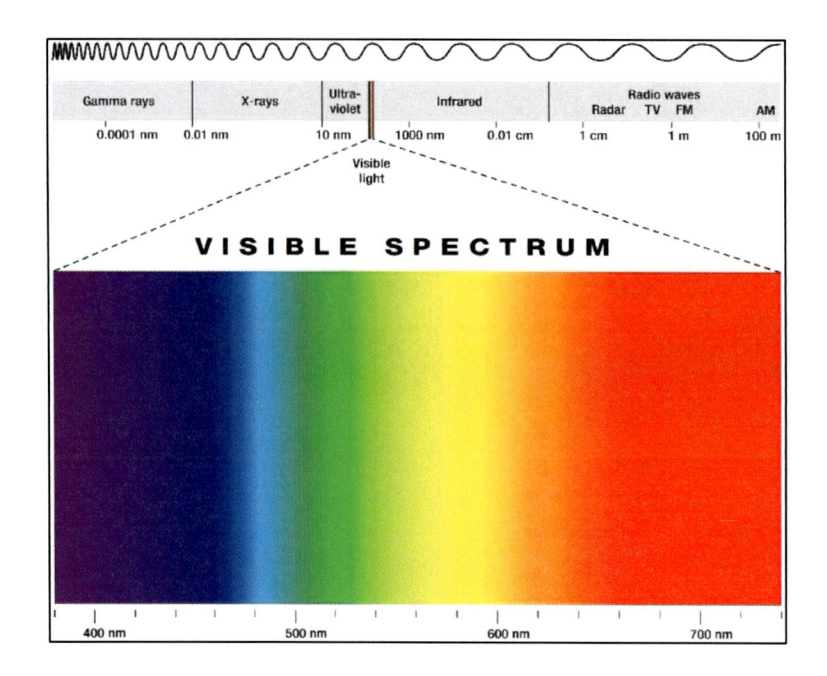

Light, because of its electromagnetic wave properties, can be absorbed, reflected, refracted, scattered, diffracted, enhanced, dispersed, and transformed. One of the counterintuitive properties of light is that it also has a particle-like nature. That discovery was made by Albert Einstein in 1903 while he was investigating thermo-photovoltaic materials and discovered the photon. A photon is a quantum of electromagnetic radiation and is the carrier for the electromagnetic force of light. The effects of this force can be observed microscopically and felt macroscopically. Because the photon has zero mass, it can travel long distances. Like all elementary particles, photons are currently best explained by quantum mechanics and exhibit wave-particle duality, exhibiting properties of both particles and waves. Highly redshifted photons from the early phases of the universe formed the cosmic microwave background that we now observe.

With such a broad range of characteristics and uses, it is remarkable that all light travels at the same speed in a vacuum. The speed of light in vacuum is one of the few physical constants in nature, with the current accepted value exactly at 299,792,458 meters per second, or about 186,282 miles per second. As Einstein proved, the square of the speed of light is the ratio of the energy to mass of all matter upon the universe, $E=mc^2$. Just these two facts alone suggest that the light of the universe is very special. While it could just be a coincidence that light shows up in very special ways in the physical universe, it certainly suggests probing into the role of light in the spiritual world.

Spiritual Light

"Remember the clear light,
The pure clear white light.
From which everything in the universe comes,
To which everything in the universe returns;
The original nature of your own mind.
The natural state of the universe unmanifest.
Let go into the clear light, trust it, merge with it.
It is your own true nature, it is home."

~Tibetan Book of the Dead

References to spiritual Light abound in all religions and philosophies. Whether in spiritual readings, ceremonies or prayers, Light is mentioned many times. Once you focus on the word "Light", it's amazing how often you will see or hear it mentioned during your spiritual practices.

The Old Testament cites "Light" over two hundred times and the New Testament over one hundred times. Other than God or Lord, it's one of the most mentioned words in the Bible. Some of the specific passages in the Judeo-Christian Bible that reference light are:

Genesis 1:3 Then God said, "Let there be light"; and there was light.

Genesis 1:15 "and let them be lights in the dome of the sky to give light upon the earth." And it was so.

Psalm 27:1 The Lord is my light and my salvation;

Psalm 118:27 The Lord is God, and he has given us light.

Mathew 5:16 …let your light shine before others, so that they may see your good works…

Luke 1:79 to give light to those who sit in darkness…

John 1:4 in him was life, and the life was the light of all people.

John 1:8 He himself was not the light, but he came to testify to the light.

John 1:9 The true light, which enlightens everyone, was coming into the world.

John 8:12 …Jesus spoke to them, saying, "I am the light of the world."

John 9:5 As long as I am in the world, I am the light of the world."

John 12:36 While you have the light, believe in the light, so that you may become children of light."

John 12:46 I have come as light into the world, so that everyone who believes in me should not remain in the darkness.

Acts 13:47 For so the Lord has commanded us, saying, "I have set you to be a light for the Gentiles, so that you may bring salvation to the ends of the earth."

Acts 22:6 …a great light from the sky suddenly shone around me.

Ephesians 3:9 …and to bring to light [for all] what is the plan of the mystery hidden from ages past in God who created all things…

Ephesians 5:8 … now you are light in the Lord. Live as children of light…

James 1:17 …all good giving and every perfect gift is from above, coming down from the Father of lights…

1 John 1:5 God is light, and in him there is no darkness at all.

1 John 1:7 But if we walk in the light as he is in the light, then we have fellowship with one another…

Revelation 22:5 Night will be no more, nor will they need light from lamp or sun, for the Lord God shall give them light…

[New Revised Standard Edition (1989) Christian Education of the National Council of the Churches of Christ in the US]

As seen above, Christians see God as Light, the Light of the World, the true Light, the Father of Light and the Lord of Light.

Light holds great significance as the guidance for believers of Islam. God the Almighty revealed the holy Qur'an to his beloved final prophet Muhammed. "Allah guides his Light to whom He wills and Allah sets forth parables for mankind, and Allah is knowing of all things." Sura 24 of the Qur'an (Q24:35) contains the Verse of Light: "Allah is the Light of the Heavens and Earth, the likeness of His light is as a niche wherein is a lamp (the lamp in a glass, the glass as it were a glittering star) kindled from a Blessed Tree".

In Hinduism, Diwali, the Festival of Lights, is a celebration of the victory of light over darkness. A mantra in Brhadaranyaka Upanishad (1.3.28) urges God "to from darkness, lead us unto Light". The Rig Veda includes nearly two dozen hymns to the dawn and its goddess, Ushas.

Buddhist scripture speaks of numerous buddhas of light, including a Buddha of Boundless Light, a Buddha of Unimpeded Light, and Buddhas of Unopposed Light, of Pure Light, of Incomparable Light and of Unceasing Light.

Light is also the core concept in Iranian mysticism. The main roots of this thought is in the beliefs which define The Supreme God as the source of light. This very essential attribute is manifested in various schools of thought in the Middle East.

Manichaeism, the most widespread Western religion prior to Christianity, was based on the belief that god is pure light. Manichaeism's legacy is the word manichaean, relating to a dualistic view of our world, dividing things into good or evil and light or dark.

Possibilities

Not surprisingly, physical light is an essential element of spirituality. Before modern religions, like Judaism, Christianity and Islam became widespread, early philosophies and beliefs naturally connected the light from the sun (and its reflection by the moon) to supernatural beings and powers. What is interesting is that light continues to be prominent in modern religions, even more so than in the past.

Simultaneously, science has made great advances in understanding universal light, galactic light, electromagnetic light, visual light and quantum light. While past religions likely believed that the sun, moon, lightning and fire were mystical, modern religions attribute most, if not all light, to God.

If many religions believe that God is both the Source of Light and Light itself and science has proven that light is the fundamental of element at both the microscopic and macroscopic levels of the universe, then everything in the universe is Light. While "God" may be indescribable, God as Light may be much more understandable. If believers see Light (or GodLight) as the fundamental element of everything in the universe, then understanding, collaboration, mutuality, respect and universal love may be a more reasonable possibility.

SpaceTime/Space & Time

SpaceTime

In science, a model that has been verified with a significant amount of real data and that consistently correctly predicts the outcome or behavior of an event, is usually accepted as valid. SpaceTime is such a model which fuses the three dimensions of space and the one dimension of time into a single four dimensional geometry.

Until the 20th century, it was assumed that the three-dimensional geometry of the universe, that is, how we express an object or place in terms of coordinates and distances, was independent of one-dimensional time. Prior to Einstein's pioneering work on relativity, scientists had two separate theories to explain physical phenomena: Newton's laws of physics which described the motion of objects and Maxwell's electromagnetic theory that explained the properties of light.

In 1905, Einstein postulated that the logical consequence of combining the laws of physics and the equations of light results in the inseparability of the four dimensions of space and time that were previously assumed as independent. This four dimensional model, called SpaceTime, has changed how we look at our world and the universe. Besides the dimensional implications of the model, the model confirms many other counterintuitive findings. For example, the speed of light is a constant, regardless of the frame of reference in which it is measured. But the distances and time ordering of pairs of events change when measured in different frames of reference.

While the discovery of SpaceTime by Einstein changed the world of physics, it also changed the way that the universe is viewed. Basically, on Earth we live in three dimensions with time being a variable. But the universe, and every event in it, has four dimensions. In the universe, an event is the instantaneous physical situation or occurrence associated with a point in SpaceTime (a specific place and time). For example, something as profound as the conception of a child is an event; it occurs at a unique place and a unique time. But something simple like the shattering of a vase is also an event. SpaceTime is where events in the universe take place.

The universe appears to be a smooth SpaceTime continuum, but events such as matter and energy bend SpaceTime and the geometry of SpaceTime forces matter and energy to behave in a certain way. The SpaceTime of the universe is usually interpreted from a geometrical perspective, with time as the fourth dimension. By combining space and time into a single spatial dimension, physicists have been able to understand and describe the workings of the universe in a more unified way.

Space and Time

On Earth, it has become very productive to characterize an object by its space dimensions at a particular time. For example, the position (x,y,z) of a particular sight or city on a map or a spherical reproduction of the Earth can be very precise. But the space dimensions of any object change with time. The globe of the Earth created in 1910 looks very different from a modern globe depicting the countries of our world in 2010. Old maps of our world are interesting from a historical perspective, but they are not helpful for planning next month's trip. Once a second or a minute or an hour goes by, everything changes. If the time change is small, we may not notice that everything has shifted. But it has and the present is different from the past and the future will be different from the present.

Religious stories and beliefs are perfect examples of the separation of these space and time dimensions. Pilgrimages by Christians, Jews, and Muslims to the Holy City of Jerusalem are special events to connect the pilgrims to the past foundations of their religions. While the events of 2000 years ago will not actually repeat themselves during the pilgrimage, the pilgrims relive the past by following the paths of the Holy Men of their religion.

If they could step back in time, they would likely want to be part of what happened years ago. But the arrow of time is always forward into the future and not back into the past. We haven't yet figured out how to insert ourselves into the actual past. Instead, religions and philosophies use stories and accounts of the past to connect their believers to the life stories of their spiritual heroes. Even with all the current technology, the best we can do is capture a present event on camera and then show the film after the event has passed. While it works to help people understand the past, it's not the actual event. We will undoubtedly continue to do this, but it's not the real thing. The only way to really understand what happened 2000 years ago in Jerusalem, on the Ganges and in the mountains of Tibet is to have been there.

Possibilities

Because every event that occurs in the universe is captured in SpaceTime, it can be viewed as repository of all past events that have occurred. Once other events occur, either in the present or the future, they are also captured in SpaceTime. Since the universe is a SpaceTime continuum, the universe itself contains everything that has occurred in the past and will occur in the future. From a human standpoint, every moment of our lives is captured in the dimensions of SpaceTime. After we die, our lives are still in SpaceTime, but in the past. We believe that death ends our activity in the world. But it doesn't because the history of our lives in x, y, z, and t coordinates is still available. While this is physically correct, can it influence our spiritual beliefs and the relationship of our lives to past and future events?

Muslims, Christians, Jews, Buddhists and Hindus believe that the way they live their lives has a major impact on what happens to their souls after they die. Souls are usually described as the part of a person that consists of their mind, their character, their thoughts and their feelings. Many religious and spiritual people believe that their soul continues to exist after their body is dead. Buddhists and Hindus believe in reincarnation while Christians and Muslims believe in a life after death, hopefully in heaven. If the SpaceTime of a person could be examined, great insights could be obtained. If a person's SpaceTime exists before they are physically born, it would indicate a pre-life. Similarly a post-life would see an extension of their SpaceTime connected to a person after they physically die. If the person's SpaceTime extends into a very-altered but positive state, this could indicate that their soul has moved to Heaven, conversely with Hell. Or there might be part of SpaceTime devoted to Heaven or Hell.

Dark Matter/Mystery

Dark Matter

The universe is mainly composed of dark matter, dark energy and ordinary matter. The universe also contains electromagnetic radiation, which constitutes less than 0.01% of the total mass-energy of the universe, as well as antimatter. The percentages of all types of matter and energy have changed over the history of the universe. For example, the total amount of electromagnetic radiation generated within the universe decreased by one-half in the past 2 billion years. Today ordinary matter, which includes all of the galaxies, stars and planets of the universe, accounts for only 5% of the contents of the universe.

Dark matter, a mysterious form of matter that has not yet been identified, accounts for about 26% of the mass of the universe. Dark energy, which is the energy of empty space and is the cause of the acceleration of the expansion of the universe, accounts for the remaining 69%.

Ordinary matter tends to clump into larger forms. That is, cosmic atoms condense into stars, most stars into galaxies, and most galaxies into clusters and superclusters. Since visible stars, gas inside galaxies and clusters account for less than 10% of the ordinary matter contribution to the mass-energy density of the universe, a great majority of ordinary matter in the universe is unseen. However, stars produce nearly all of the light that we see from Earth.

Dark matter is invisible to the entire electromagnetic spectrum but it accounts for most of the matter in the universe. The existence and properties of dark matter are inferred from its gravitational effects on visible matter and its radiation through the universe. Dark matter has not been detected directly, making it one of the great mysteries in modern physics. Dark matter neither emits nor absorbs light or any other electromagnetic radiation at any significant level.

Dark energy, an even greater enigma than dark matter, is an unknown form of energy that is hypothesized to permeate space. The density of dark energy is much less than the density of ordinary matter or dark matter but it dominates the mass/energy spectrum of the universe.

In summary, 95% of the mass-energy of the universe is unknown or dark and less than one-tenth of the remaining 5% of ordinary matter is observable. The observable physical world, which we often treat with great certainty, amounts to less than 0.03% of the universe, the other 99.7% is a mystery.

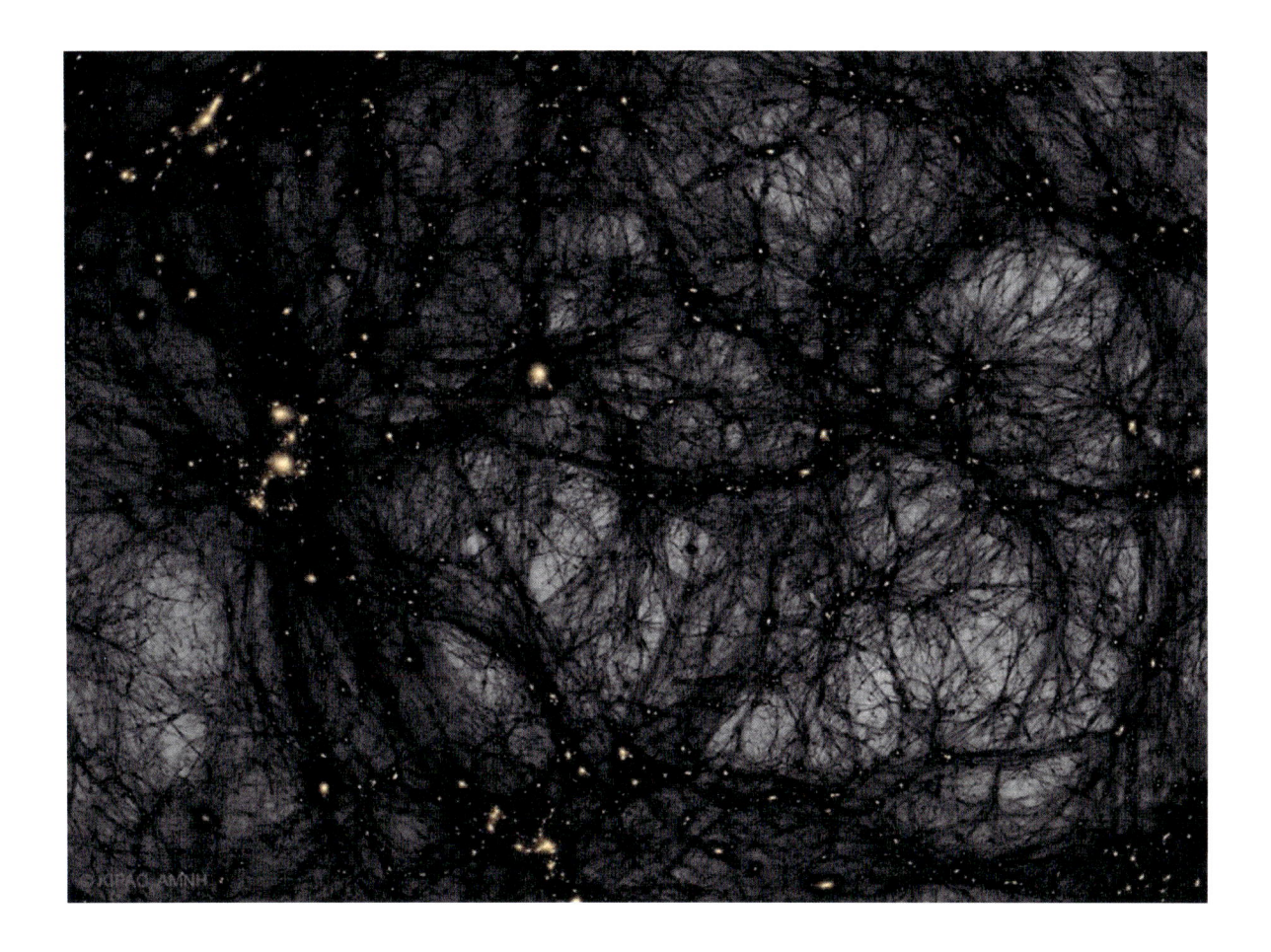

Mystery

Spiritual and religious beliefs have always been full of mystery. Almost everything was mysterious to early humans. Eventually, Homo Sapiens figured out that certain things, like the seasons, the tides, the movements of the sun and moon, and even the vulnerabilities of mammoth animals, were somewhat predictable. While birth and the inevitability of death likely were understood early on, exactly why life and death happened and what might occur before birth and after death was the focus of many early belief systems. Egyptian, Roman, Greek, and Middle Eastern religions often assigned these mysteries to the workings of gods. While the development of Judaism, Christianity, and Islam consolidated the gods into one God, God is still the great mystery of most modern religions.

As the major theistic religions matured, God became more defined. These religions have sacred writings, like the Bible and the Qur'an, that characterize, describe and personify their God. Nevertheless, there is still many unknowns in all of the current philosophies and religions about the nature of God, the universe, the afterlife, and even the role of human beings on the Earth. Certainly, the great advances that have been made in physics, economics, and medicine have improved the quality and longevity of life for many people in our world. Life is more predictable and manageable than ever before. But death, particularly what happens after death, is still the great unknown. All of the science in our world has still not given us any clues about the afterlife. On Earth, the arrow of time points in one direction and the knowledge of the future is not available to us. Occasionally, anecdotal descriptions of what happens after death are reported. Interestingly, most involve the appearance of a great light and a beckoning for the soul to leave the body and transcend to heaven or God. Death, while uncertain to even the most devout believer, is less frightening when you believe that you will be transported from the dark into the light.

Possibilities

Could it be that, like the universe, the future is mostly dark? As the universe expands and time moves forward, what was the future but is now the present, becomes brighter. At any moment, the future is unknown but the present is in the light. The present is the edge of the expanding universe. Beyond that, everything is unknown and mysterious, therefore dark!

But as the universe expands, it will become better known and should provide us with new information, knowledge and wisdom. Rather than being fearful or anxious about the future, it is our destiny. How we act in the present has a great effect on the future. If we see our lives more as influencing the future rather than treating the future as mysterious and potentially ominous, everything changes. Holding both the present and the future in our hands is nothing really new, but living that way could be very powerful.

If humanity truly believed that they, and not some mysterious force, are responsible for how the future turned out, what a difference that might make. The possibility of mankind owning the future could change everything. Some organizations carefully develop a 100 year strategic plan to guide their present actions and goals. Is a 10,000 year strategic plan for our world a possibility?

Black Holes/Hell

Black Holes

One of the most interesting phenomena in the universe, black holes, are regions of SpaceTime where gravity is so strong that nothing, no particles or even electromagnetic radiation such as light, can escape. The theory of general relativity predicts that a compact mass can deform SpaceTime to form a black hole.

Although it has an enormous effect on the fate and circumstances of an object crossing it, according to general relativity, it has no locally detectable features. In many ways, a black hole acts like an ideal black body as it reflects no light, making it essentially impossible to observe directly.

Black holes of stellar mass form when very massive stars collapse at the end of their life cycle. After a black hole has formed, it can continue to grow by absorbing mass from its surroundings. By absorbing other stars and merging with other black holes, supermassive black holes of millions of solar masses may form. Stars passing too close to a supermassive black hole can be shred into streamers that shine very brightly before being "swallowed."

Hell

There's a Hell in every religion. Whether it's torture, isolation, or separation, Hell is the opposite of Heaven. It's the place where bad people go for eternity, unless they repent before they die. It's also a great deterrent to bad behavior while we are alive.

According to Christian culture, heaven and hell are essentially deserved compensations for the kind of Earthly lives we live. Good people go to heaven as a reward for a virtuous life, and bad people go to hell as a just punishment for an immoral life. In that way, the scales of justice are thought to be in balance. But there is a wide variety of views about Hell, including very different conceptions of divine love, divine justice, divine grace and free will, along with its role in determining a person's ultimate destiny and very different understandings of moral evil.

In Islam, Hell (Jahannam) is an afterlife place of punishment for evildoers. The punishments are carried out in accordance with the degree of sin that a person has committed during their life.

Naraka is the Hindu (and Buddhist) equivalent of Hell, where sinners are tormented after death. It is also the abode of Yama, the god of Death. The Hindu religion regards Hell not as a place of lasting permanence, but as an alternate domain from which an individual can return to the present world after crimes in the previous life have been atoned. These crimes are eventually nullified through an equal punishment in the next life.

In summary, all religions believe that Hell, regardless of the details that each religion uses to describe this negative afterlife, is very undesirable. It's generally described as dark, void of light and difficult or impossible to leave once you're there.

Possibilities

The physical description of BlackHoles, that is, places in the universe where there is absolutely no light and from which no light can escape, is certainly similar to the spiritual descriptions of Hell. If God is any form of light, then, like Hell, a BlackHole is no place for God.

Since SpaceTime contains everything in the universe, we would expect to find Hell in SpaceTime and a BlackHole seems to be a reasonable candidate for Hell. The lack of light inside a BlackHole may be the important insight for all of us who are trying to avoid Hell. We might avoid an afterlife of Hell by being and living in the Light when we are alive and creating our own SpaceTime. As with physical dynamics, even coming close to a BlackHole or Hell is extremely dangerous since, once under the influence of this dark place, escape may be hopeless. While this may simply reinforce what most of us already believe about staying in God's Light, the vision of being trapped in a BlackHole forever could certainly motivate us to avoid those dark places and temptations in our lives.

The Universe/Heaven

The Universe

At present, the universe continues to expand, as it has since the beginning of time and space. There is no indication that the universe will not continue to expand. If so, SpaceTime will continue to increase and the universe may have no end. In other words, it may exist forever and ever. A universe that never ends is hard to fathom but certainly reasonable since the universe has existed for 14 billion years.

Nevertheless, cosmologists have considered two possible scenarios for the end of the universe. The universe could reach a maximum size and then begin to collapse. It would become denser and hotter again, ending with a state similar to that in which it started, a Big Crunch. Alternatively, the expansion could slow down causing stars to burn out, leaving behind white dwarfs and black holes. Collisions between these remaining objects would result in mass accumulation into larger and larger black holes. The average temperature of the universe would asymptotically approach zero, a Big Freeze. While the end of the universe is a possibility, it does not seem consistent with the creation and accelerating expansion of the universe. In the above scenarios, the universe ends up either as it started, an infinitely energetic singularity, or nothing at all.

Regardless of the fate of the universe, the future of planet Earth is much more predictable.

Assuming human beings do not destroy this planet, the likelihood of planet Earth existing for many more millions of years is very slim. The Earth could be hit by a meteor, our sun could burn out, our galaxy could collapse into a black hole, or the planet could be swallowed up into a supergalaxy with the destruction of our solar system. If any of these possibilities occur, our universe could go on, SpaceTime would increase, and we would be history in a past SpaceTime. The totality of SpaceTime would include all of our lives and, if the universe does not end, would exist forever.

Heaven

Christians, Jews, and Muslims believe that there is an afterlife. And every believer hopes that after they die, they end up in Heaven. The Roman Catholic Church teaches that after death there is a state of Purgatory. This is a place where some people who have sinned are purified in a 'cleansing fire', after which they are accepted into Heaven. Christians who have lived a good life may pass directly to Heaven. Really bad Christians go to hell, perhaps for eternity.

Hindus and Buddhists believe in a form of the Christian Heaven where souls go for a time but may come back to Earth either at a higher level of consciousness or a lower level, sometimes as animals. As the reincarnated Hindu and Buddhist returns to Earth, the goal is to live a better life and through various reincarnations to achieve enlightenment and then Nirvana, a place of perfect peace and happiness.

Possibilities

Since everyone creates their SpaceTime signature as they live their lives, these signatures exist after we die. Depending on when you believe the human race (with souls) occurred, SpaceTime is filled with many former human SpaceTime experiences. As men and women continue to live and die on Earth, more Human SpaceTime experiences will be added. Whether or not this is the final "resting place" for human beings, SpaceTime experiences may determine if there is something more for the souls of human beings than just being in past Spacetime. If the Earth were to cease at some point, no additional experiences would occur but SpaceTime would continue. Perhaps the end of the Earth will trigger a transformation in it's SpaceTime to another state. That state could be Heaven and the waiting time could be Purgatory.

Overall, most people hold onto the hope that they will join God after they die and that they will reside with God for eternity, in a place called Heaven. Perhaps it's as simple a possibility as this, "You will live your life on Earth while creating your SpaceTime signature. When you die, your SpaceTime will be with God in his universe forever. Your Light will be one with GodLight for eternity". Of course it's likely that some religions would add that you just might be gobbled up by a Black Hole and that would be Hell!

Infinity/Eternity

Infinity

In science, infinity is a term that comes up primarily in mathematics. When a number, any number, is divided by zero, the result is the term "infinity". When that happens, it's usually the end of the mathematics problem because infinity cannot be described or acted upon and the term has no boundaries. Anything divided by zero is unimaginably large. So "infinite" is now used to describe time, space, velocity and other physical parameters when they are unimaginably large.

While most people think that the universe is infinitely large and will exist for an infinitely long time, that may not be true. The most compelling argument against an infinite universe is that we know that our universe started as a very small point about 14 billion years ago. While we may think that the starting point, the Big Bang, had an infinite mass in order to create our universe that appears infinitely large, it did not. So while the concept of infinity may be useful in mathematics, it is not very useful for much else. After all, what could be bigger than the universe?

Generally, infinity is used to describe something that we believe will go on forever or have dimensions that never end. While we may not have concrete examples of these (since the universe doesn't qualify), it's both scientifically and spiritually useful to convey endless time or dimension. To say that God is infinite or endless may not be provable, but it certainly suggests that God has unique characteristics. On the other hand, how do we really know?

Eternity

Many religions use the word "eternity" to convey a forever and neverending quality, mostly applied to time. Eternity is used in many spiritual writings to convey infinite time that never ends. Often God is described as eternal, meaning that God has existed and will exist forever, without beginning or end. More personally, eternal life as applied to a person's soul, is a cornerstone belief of many religions including Christianity, Judaism, Islam and Hinduism. Conversely, Buddhism teaches the very opposite, the impermanence of everything including the soul. Eternity is a concept that applies to things that cannot be counted or measured.

Possibilities

While infinity and eternity are often used in the same context, this may be very misleading. Infinity really describes a very large number. Anything divided by zero produces infinity. One could imagine the beginning of our universe as a product of something very small, as is true in the first nanoseconds of the Big Bang, divided by nothing.

Other theories of the cause of the universe point to a separation of "something" into two parts, positive and negative. Whether our universe is the positive or the negative result of this separation is unknown. But either the positive or the negative "universe" could have the energy necessary to expand into a universe similar to what we see as our universe. Since anything divided by zero or created from zero could be infinite, the creation of an infinite universe could have been created from nothing. But why would such a thing occur? This may be where a Creator God comes in.

Because we now appreciate the massive size of the universe, it's easy to think of our universe as infinite. However, we know that it did not begin an infinitely long time ago. The age of the universe is finite, somewhere around fourteen billion years. While our mind cannot fathom such a long period of time, the universe, as of now, is finite. When it will end nobody knows and there are many theories that suggest that the universe will not go on forever. So the jury is out as to our universe being infinite.

Eternity, on the other hand, is somewhat easier to understand. Eternity doesn't have to be linear, like infinity. Given the proper energy, anything traveling in a circular orbit that is not slowed down by friction, will theoretically, go on for eternity. Many things in the universe, such as galaxies, stars, planets and moons, are balanced by centrifugal and gravitational forces that keep them orbiting in a stable manner. We know that this doesn't go on forever because stars run out of energy and planets are affected by other bodies. However, the general state of the universe seems to favor systems that could go on forever.

Finally, eternity, unlike infinity, is more qualitative than quantitative. We pray to an eternal God, who we believe has been and will be around forever. We assume the universe is eternal, but realize our lives are very finite. The distinction between infinite and eternal is very human, more connected to forever than long or large. We yearn to be eternal with God, but what does that really mean? Thinking about eternity from an infinite perspective may not be very useful. Eternity could simply be believing that our place in the universe will not end. While our physicality will certainly end, our impact on the universe could go on forever, simply connected to the eternal possibility of our universe.

Planets/Earth

Planets

Science taught us that the planets in our solar system started as high-velocity masses sailing through the universe that were captured by the gravitational pull of our sun. Under the right conditions, these big rocks were forced into elliptical orbits that have been rotating around our sun for billions of years. We now know that there are millions, if not billions, of planets rotating around different suns in different galaxies in our huge universe. Earth happens to be in a particularly fortunate orbit around our sun that provides excellent conditions for the possibility of life. Our solar planetary neighbors experience different conditions which may or may not have the possibility of life. Nevertheless, planets around suns are a normal part of the universe.

Since we do not have a good view of the entire universe, there may be millions of planets that have an atmosphere and the energy necessary to replicate our Earthly planet. From that perspective, we may not be unique.

This has led to beliefs that extra-terrestrial or alien life does exist. However, extensive scientific examination of reported incidents of alien beings that have traveled to our Earth has not supported the validity of these reports. There may be other planets that have the properties of Earth but no evidence exists to prove that point.

From what we know at this time, our amazing Earthly planet is just following the laws of the universe. While we may be unique, the probability of a planet developing as it has on Earth may be small but not zero. Most scientists believe this and take this as scientific truth.

The implications for the reverse, that is a zero probability for a planet like the current Earth to exist, would presume some special divine intervention on our particular planet. Most spiritual readings suggest such a situation, that is, that God specifically chose our planet to build his Earthly kingdom.

Earth

Not surprisingly, we know a lot more about the Earth than any other planet in the universe. According to radiometric dating, Earth was formed over 4.5 billion years ago. Because of its atmosphere of nitrogen and oxygen, its distance from the sun and its orbit of 365 days around the sun, life became possible if not probable.

Within the first billion years of the Earth's history, life appeared in the oceans and began to affect the Earth's atmosphere, which then led to the proliferation of aerobic organisms. Geological evidence indicates that life may have arisen as early as 4.1 billion years ago.

Since then, the combination of Earth's distance from the Sun, physical properties, and geological history have allowed life to evolve and thrive. In the history of life on Earth, living organisms have gone through long periods of expansion, occasionally punctuated by mass extinctions. More than 99% of all species that ever lived on Earth are now extinct. Fortunately, humans have survived and, good or bad, now dominate the planet.

While we take for granted that the Earth was made for human life, this only occurred recently in the history of the planet. Our first human ancestors appeared about six million years ago, probably when some apelike creatures in Africa began to walk habitually on two legs. Later, some of these pre-historic humans spread from Africa into Asia and Europe. From then on, humans developed into what they are today, dominating the planet and very much focused on themselves and their well-being.

Today, Earth is a planet of eight billion human beings who continue to grow and use the resources of the Earth at an amazing pace. While there are constant reminders that we are living on a very small planet orbiting a very small sun in an unremarkable galaxy that is subject to all the same physical laws that govern the entire universe, we believe that we are very special.

Possibilities

From my perspective, there are many people who behave as if the universe was made for mankind, rather than human beings being a product of the universe. Most of our spiritual literature reinforces the idea that God, in very short order, created the universe, the galaxies, the stars and our planet so that we could live and thrive here on Earth.

Science sees the story in just the reverse, with humans emerging very late in the evolution of the universe and with an extraordinarily low probability of occurrence. Fortunately, humans did come about after the universe was 14 billion years old but only in the last million years on this tiny planet called Earth. We have been in the universe for 0.01% of the universe's history and probabilistically, very lucky to have survived on this planet. Increasingly, we are using up most of the Earth's resources and focusing more inwardly than universally.

Many of the early religious practices reinforced the notion that the Earth was a gift to human beings that carried great responsibility. The soil, the water, even the air were to be used in a caring, responsible way. In 1990, indigenous people spoke at a global conference on the environment and said, "The difficulty, as we saw it, was with the settlers themselves and their failure to tread lightly, with humility and respect, on the land. The settlers wanted to live on the land, but the host people lived with the land. Living on the land means objectifying the land and natural resources and being shortsighted concerning the future. Living with the land means respecting the natural balance."

Realizing that it took billions of years for the Earth to evolve to its current state, it seems reasonable for humans to try to keep our fragile little planet as intact as possible. Furthermore, if the recent state of our planet led to the evolution of mankind, we risk a major impact on human beings if the planet changes in any significant way. Yet, the global population's lukewarm reaction to scientifically-proven global warming, serious climate change and air/water pollution is already threatening the Earth's balance. The science is clear that the Earth now faces a significant self-imposed threat that could prove very destructive to life as we currently know it. Is there a possibility that humans will react soon enough to avoid a catastrophic crisis?

Several of the astronauts that were able to view the Earth from space have reported psychological and emotional reactions that have been attributed to an "overview effect". First to see the Earth from a "God's eye view", they reacted strongly to the perceived fragility and vulnerability of the planet. This has led to worldwide discussions and support of the overview effect, which is being accelerated by commercial flights to space and a significant increase in the number of people able to experience that effect. Almost everyone that has physically experienced the overview effect have become champions for better care of our planet. Is there a possibility for non-space travelers, guided by astronauts, to become more concerned and stimulating a stronger reaction to global warming and serious climate change?

Based on all the past cosmological evidence, the Earth has a very finite lifetime, shortened even more by the self-destructive behaviors of its population. It will go away someday, regardless of how humans behave or how we view it from a spiritual and religious perspective. All of our human history is strongly tied to the opportunities and threats provided by the planet. If we realized how transient the planet, as well as we, truly are, we might become much more spiritual, collaborative and caring. Could we think of ourselves as universal voyagers rather than inhabitants of this small planet? Would that change how we view God, our current home and our planetary neighbors?

Evolution/Creation

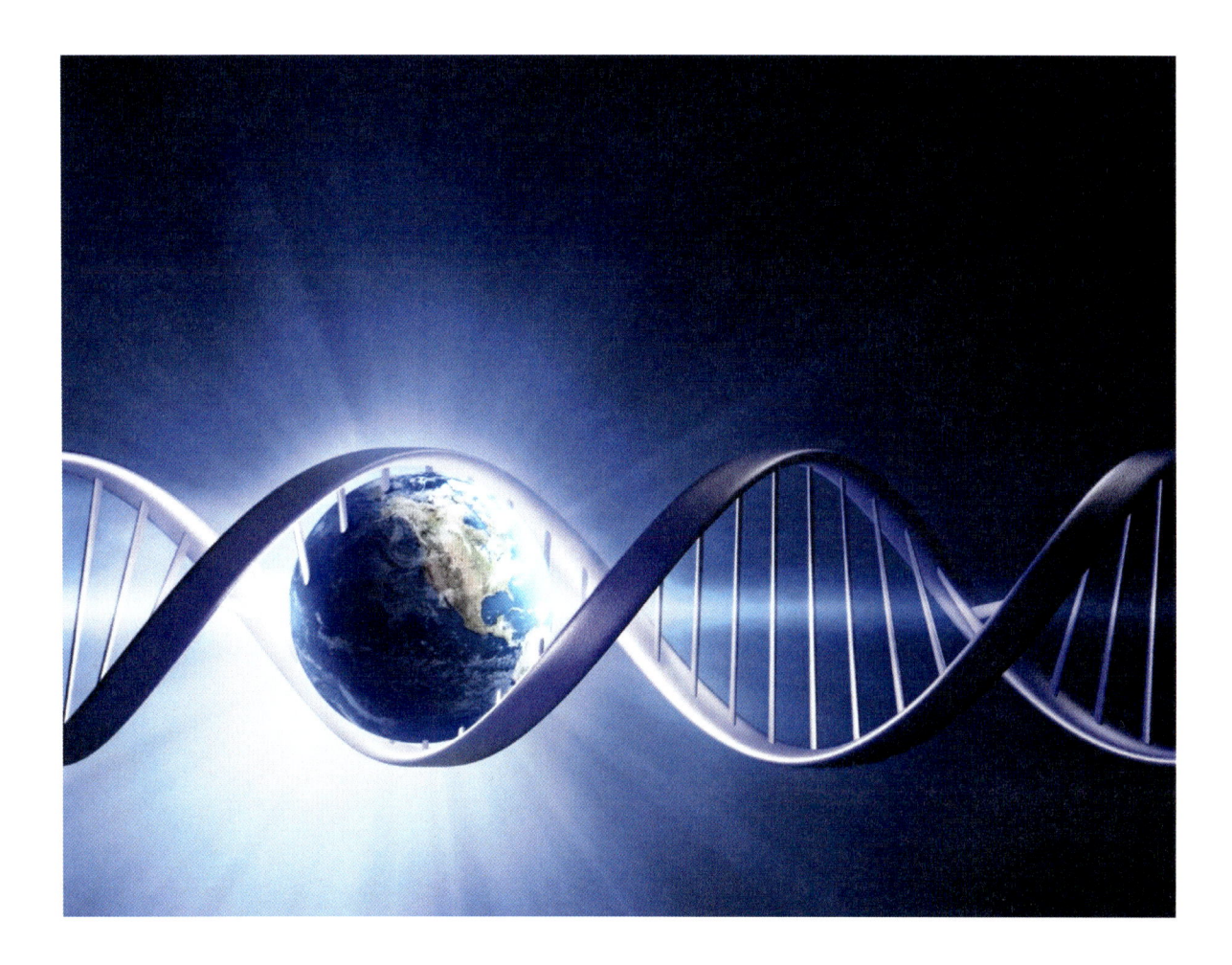

Evolution

Archeological evidence suggest that there were life forms on Earth four billion years ago. These may have been single cell organisms that were formed in the primal waters and air of our planet. While some support the theory that life came from other planets or galaxies, their is no evidence to support this.

Darwin, in his "Origin of the Species", wrote convincingly that our species started with these basic life forms and evolved through various species modifications and enhancements. Most of the modifications favored species who were better able to defend themselves against predators or more adept at finding food.

Complex life forms, like fish, snakes, and birds, took billions of years to develop. Pre-historic Homo-Sapiens, like Cro-Magnon Man, only evolved in the last 45,000 years. From about 45,000 to 10,000 years ago, the planet experienced massive changes and the evolution of Homo-Sapiens to modern man only occurred about 7,000-10,000 years ago. Most historical documents follow human-beings from about 4,000 BCE to today, a very short period of 6,000 years. Human beings today have evolved to be the most successful species on Earth at protecting themselves, destroying their enemies, finding food, harvesting food and developing extremely lavish lifestyles.

Darwin's evolutionary theory, while vigorously attacked at that time by many religious organizations, is now widely accepted as the truth. Christians, Jews, Muslims, Hindus and Buddhists accept evolution as the origin of our species. However, most religions see human evolution from animals as only one aspect of the development of modern man. Somehow, humans evolved much further with souls that would live forever, dispositions that could be much more peaceful than animals, and minds that are more pre-occupied with lifestyle than survival against animal predators. Nevertheless, modern humans still have very significant parts of their brain focused on optical processing to detect potential harmful threats, fear processing centers to prepare for any needed arousal, and quick response capabilities for when physical action is needed. We may look and act very different from our nearest species, the orangutans, but a big part of our brain is like theirs.

Creation

Just like the creation stories of the universe, all religions have human creation stories. For Christians, Jews and Muslims, it's the biblical story of Adam and Eve. Man came first, created by God out of mud, and was alone in the world. Then God realized that man needed a partner, so God took a rib from man and formed woman. Adam and Eve were perfect at that moment and they would have enjoyed eternity with God on Earth. But the devil, in the form of a serpent showed up. He tempted Eve to take a bite out of an apple from the forbidden tree and then seduced Adam to do the same. From then on, humans were tainted by this "original sin". They needed to work to survive and they were no longer immortal. Women were second class citizens and men and women had to be saved from their evil ways.

While many religious people do not take this story literally, it usually is learned at a very early age. As such, it is imprinted on our young minds and may influence how we behave as we age, either consciously or sub-consciously.

Regardless of what we believe, this creation story reinforces the idea that human beings chose to be sinful by desiring something that God did not want them to have. Not only did they disobey God, human beings were then forced to choose between good and evil for the rest of their lives because of Adam and Eve's sin.

In the Hindu human creation story, the first woman on Earth, Shatarupa, is the daughter of the creator god, Brahma. The first man on Earth is Manu, also a son of Brahma. Shatarupa married Manu and had five children. Manu gave his first daughter and middle daughter to sages and his youngest daughter back to God. While the further population of the Earth is not as clearly described as in the Judeo-Christian Bible, the belief that Manu's female children were given to sages suggests that Hindu people were originally wise and godlike.

In Buddhism, humans originated at the beginning of the current age as deva-like beings reborn from the Abhasvara deva realm. They were beings shining in their own light, capable of moving in the air, living a very long life, and not requiring sustenance. The Buddha then presents an elaborate myth in the Agganna Sutta that explains how humans became bound to the wheel of Samsara and life-after-life in the Six Realms. In some ways, the Buddhist creation story is similar to the Bible story where humans are originally godlike but then succumb to life's challenges until they can achieve Nirvana after successive reincarnations.

In summary, all religions believe that humans were created by God (or many gods) as pure and good. In various ways, the first humans succumbed to worldly challenges and sinned. After that, life was harder and death was inevitable. Only by living a better life or improving their lives in various reincarnations are they able to join God in Heaven or reach Nirvana.

Possibilities

Rather than assume that the downfall of human beings was caused by sin, selfishness, and egotism, evolution presents another possibility. Instead of believing that humans had a choice in the matter, the science of human evolution suggests that human beings evolved over long periods of time from lesser species whose main concerns were on preservation and reproduction. Preservation required aggressive behavior, whether for flight or fight. Reproduction required opportunistic and bold behavior, without much consideration for the "feelings" of the mate.

Species evolved according to the rule of "survival of the fittest" which favors ever-increasing boldness and cleverness. When pre-historic human beings finally evolved, their DNA was strongly influenced by their aggressive predecessors. With continued evolution, characteristics such as curiosity, cleverness, and creativity allowed the more sophisticated human-like species to survive. As a consequence, the current humans occupying the planet have DNA profiles that were "crafted" to be very aggressive: physically, sexually and psychologically. Rather than the result of an "original sin", human beings are the way they are because of their "origin".

Although it certainly can be argued that any bad behavior or sin is the responsibility of the "sinner", human beings seem pre-disposed to feelings and behaviors that are categorized as sinful. As challenging as it might be, the religious and spiritual goal of human beings is to achieve a sinless life, either through sacrifice, self-denial, ideal behavior or possibly, reincarnation. Given how we started, this is a tough road, made even tougher by the belief that we chose this situation before we were ever born. Perhaps by viewing humans not as tainted by the original sin of the first man and first woman but by their natural evolution, a new outlook for mankind can be established. Surely it will be difficult to achieve "world peace" or collaboration or Nirvana from either scenario. But recognizing that our human nature favors the aggressive rather than the weak might be a good start.

Humans/Souls

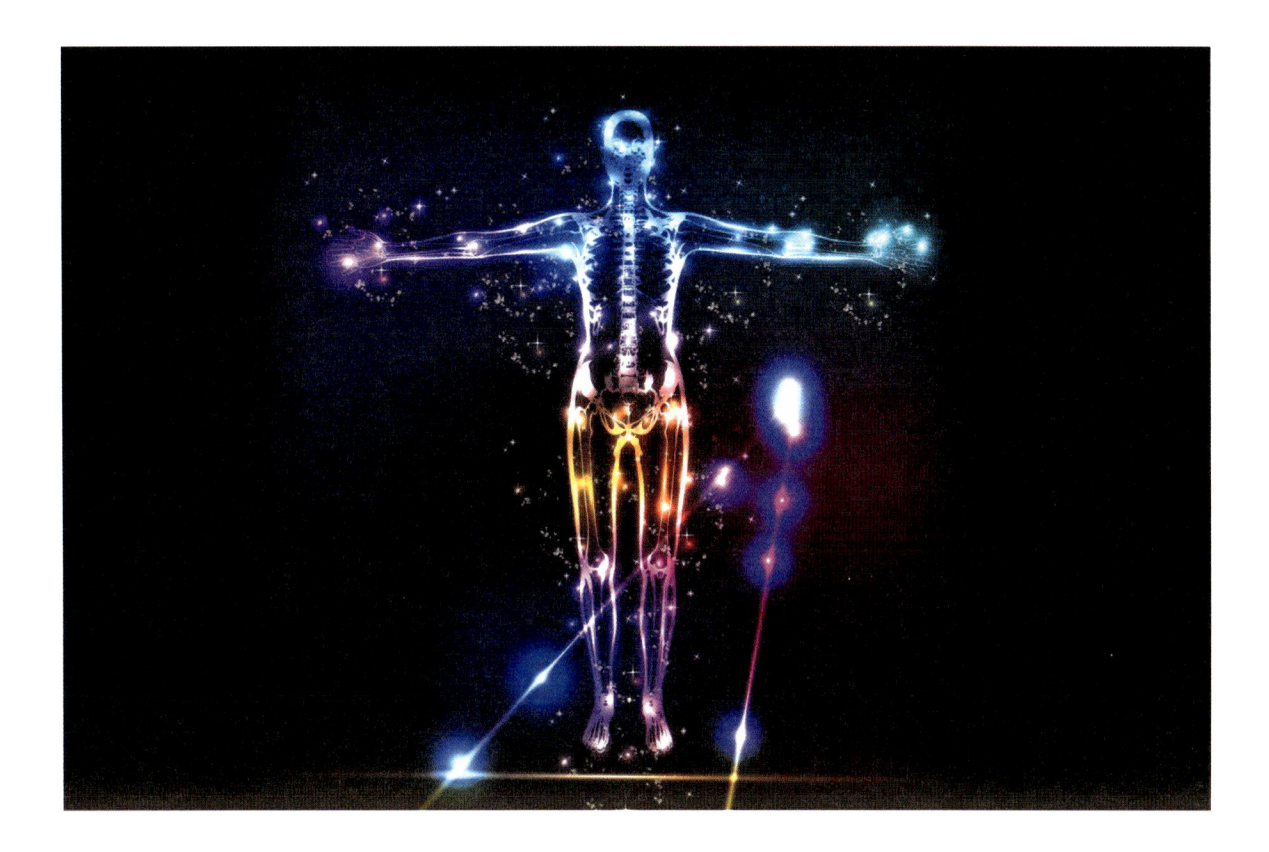

Humans

Today, humans generally live, at most, one hundred years. Many have serious illnesses long before then and some die very early. In many countries, undesirable embryos can be terminated in uterus and euthanasia of the sick and old occasionally occurs. Most human beings no longer have to protect themselves from animals to survive. Instead, humans are quite vulnerable to viruses and bacterial infections. The most dangerous predators on the planet are human beings and their destructive capabilities have been advanced over the past few centuries to catastrophic levels.

In the 6,000 years that modern humans have existed on the Earth, the scientific study of our species has progressed to the point where we can determine the exact DNA of any person. Soon we will be able to predict the DNA of a newborn child before they are conceived. Since DNA is the blueprint for the development of the resulting child, tailored babies are in the near future.

Once conceived, the growth path of the child and later the adult could be re-designed by altering the DNA throughout their life. As human beings age, further DNA changes could prolong their lives, possibly for a very long time. Theoretically, but not likely, a "perfect" human could be produced that would last forever. This may never happen, but we, as humans are moving in that direction. As this happens, the nature of man would definitely change, hopefully for the better. Or for the worse, if only a few super-humans dominate the planet to the detriment of all others of lesser capabilities.

The gender, race, and characteristics of human beings are also in major flux. Today, a person can choose between more than a dozen terms to define their gender, independent of their physicality at birth. Cross ethnicity marriages are growing in every part of our world and the differences between people based on place of birth or color of skin is no longer distinct. After DNA modification, designer children are now growing to unusual heights and weights with enhanced immune capabilities.

At the current rate of progress, the traditional conception of a human child through the fertilization of a female egg by male sperm could be bypassed by manipulation of DNA from non-human sources. At some point, who or what we consider to be human beings could change dramatically. If that happens, what about our souls?

Souls

The soul of a person has been described in many ways. In philosophy, a soul is that immaterial essence that confers individuality and humanity on that person. In psychology, the soul is often considered to be synonymous with the mind or the self. In theology, the soul is defined as that part of the person which partakes of divinity and often considered to survive the death of the body.

Many cultures have recognized some incorporeal aspect of human life or existence corresponding to the soul and have attributed souls to all living things. There is evidence that prehistoric people believed that the soul is distinct from the body but resides in it. The ancient Chinese distinguished between a lower soul that disappears with death and a rational soul which survives death.

Biblical references to the souls do not distinguish between the ethereal soul and the corporeal body. Christian concepts of body and soul separation were introduced by Saint Augustine who wrote of the soul as a "rider" on the body. St Thomas Aquinas referred to the soul as the motivating principle of the body, independent but requiring a body to make a person. The early

Christian philosophers adopted the concept of the soul's immortality and believed that the soul was created by God and infused into the body at conception.

In Hinduism, the soul or atman is the universal and external self of which each individual soul partakes. At birth, the soul is imprisoned in the Earthly body and at death, the soul passes to a new existence determined by karma.

The Muslim concept of the soul, like that of Christians, holds that the soul comes into existence at the same time as the body and leaves the body at death.

Overall, most believe that the soul is part of the person, but separate from the body. How the soul changes during the lifetime of a person is generally not discussed. At conception, the soul enters the body through infusion from God. At death, the soul leaves the body and either goes to heaven or hell or reincarnates depending on the karma of the person. The soul of a person is connected to that person but the soul is also connected to God, before, during and after that person's life.

Possibilities

Could our souls be GodLight that we receive at conception? At that moment, we are probably closer to God than we will ever be. As the embryo grows in the uterus, our souls grow with it. At birth, we join the outside world both physically and spiritually. Then we are subject to all the temptations and corporal needs of the human body.

Babies certainly seem to be the most perfect phase of human development, innocent and incapable of committing sin. But babies turn into children and later adults, who certainly can err and be less than perfect. These growing humans carry their soul with them and their souls are likely affected by how they live their life. If our souls begin with GodLight, then the soul should be the constant connection with God as we live our lives. While our physicality may lead us astray from living a life that is consistent with GodLight, our souls could be our salvation. The GodLight that is part of our souls might be the light that leads us to God and, after death, living with God for eternity in Heaven.

Opportunities

We have come full-circle, from the GodLight that created the universe to the GodLight in our souls. By looking at both the physical and the spiritual aspects of our universe, our planet and ourselves, new possibilities arise. And these could lead to new opportunities.

§ Could the light and energy that created the universe be all that exists? Could that Godlight be the basis of everything in the universe, including humans?

§ Is there an opportunity to ignore the distinctions between the various aspects of the universe, such as light, energy, galaxies, suns, planets and human beings and consider them as one? Does this present more opportunities for us to see ourselves beyond our short lives and as GodLight that will live forever? Could this commonality change the way that we treat each other? Can we see each other more as the same rather than different?

§ If God created the universe, then everything in our universe is connected to God? Is there an opportunity for mankind to see their role as God's travelers in the universe? Could human beings identify more with the universe and God, than with the transient issues of our planet?

§ Could we agree that God created the universe? We know that something happened 14 billion years ago to cause our universe to begin. While we may never know the specifics of that cause, could we describe that cause as God? If so, could we see that God as our creator, our father, our origin? Could we begin with such an agreement to help us understand everything?

§ Is there an opportunity to examine light from a spiritual perspective rather than simply a physical entity? Could that lead to a further understanding of our universe and ourselves?

§ Could SpaceTime be the missing link between life, death and after-life? There is scientific evidence that everything which happens in the universe is recorded in SpaceTime. Is SpaceTime our individual or collective past, present as well as future?

§ Is dark matter God's way of showing us that, despite our self-proclaimed brilliance, we know so very little? Should we continue to try uncover the mysteries of the universe realizing that, since the beginning of time, the universe is mostly dark and unknown?

§ Are Black Holes a metaphor for Hell? Their lack of all light seems such a contrast to God as Light. Can we learn something spiritual from Black Holes?

§ Is Heaven contained in the universe? Would that hypothesis help bring human beings closer to God?

§ Is Eternity linear or curved? Can we look at this from a spiritual perspective? Can we learn from religions who believe it is not linear?

§ Is the Earth really that special? Are there more planets with living beings? If so, are we that special? Is there an opportunity to explore these possibilities?

§ Did humans evolve from earlier species? Given that is true, what opportunities are there for us to grow and evolve? Are humans generally in denial of their origins?

§ Most humans believe that they have souls. What if we could behave and interact as souls instead of human bodies? If God views us as souls, can we focus more on our spirituality than our physicality? Are there ways for humans to collaborate and communicate from a "souls" perspective?

Epilogue

> God, let me
> - Absorb
> - Reflect
> - Understand and
> - Be Your Light
>
> Let me spread your Light and Love throughout the world

Since I started writing this book, I begin every day with this simple prayer, "God, let me absorb, reflect, understand and be your Light. Let me spread your Light and Love throughout our world". It reminds me how much God loves us and how God made everything, the universe as well as us, from Light. It also helps me look at others as God's Light and I hope that makes me love them more.

About the Author

Dr. Bart Barthelemy is the Founding Director of the Wright Brothers Institute and the President of the Collaborative Innovation Institute. Dr. Barthelemy was the National Director of the National Aerospace Plane Program, where he reported to the White House and was responsible for the development of the nation's hypersonic aerospace plane. While a member of the Federal Senior Executive Service, he served as the Technical Director of the Air Force Wright Aeronautical Laboratories at Wright-Patterson Air Force Base in Dayton, Ohio, the Air Force's largest research and development complex. He has been a consultant to a variety of aerospace industry companies and federal government organizations, including Lockheed-Martin, Boeing, the Department of Defense and the Air Force Research Laboratory. Bart was also a Visiting Scientist at the Software Engineering Institute and Carnegie Mellon University and Adjunct Professor at the University of Dayton. Bart's educational background includes a Bachelor of Science in Chemical Engineering from MIT, Master of Science in Nuclear Engineering and Physics from MIT, and a Doctor of Philosophy in Nuclear Physics/Mechanical Engineering from The Ohio State University.

Educated with degrees from MIT, Dr. Barthelemy began his career as a rocket scientist and nuclear physicist. When he entered the United State Air Force, he was assigned to the Air Force Research Laboratory at Wright-Patterson Air Force Base in Dayton, Ohio. His first few years with the Air Force were as a scientist, but he was promoted to a manager at the age of 25. From there, Dr. Barthelemy took on greater and greater leadership roles and responsibilities. He became an Executive Director of the Laboratory in 1984. In that position, he managed and directed over 10,000 S&Es to develop high-end technology for the Air Force. Dr. Barthelemy retired from the Air Force in 1995 and started several companies and a non-profit called the Wright Brothers Institute (WBI). Now in his 80s, Dr. Barthelemy still serves as Founding Director at WBI and works every day on innovation and collaboration projects.

Other Books by Bart Barthelemy

High Performance, A book on high performance technology leadership (Self-published, 1985)
The Sky Is Not The Limit: Breakthrough Leadership (St. Lucie Press, 1993)
Collaborative Innovation, Wright Brothers Institute, The future of innovation, (Balboa Press in 2020)

Made in the USA
Monee, IL
01 September 2022